Living Green

Reusing

By Meg Gaertner

I0876987

level 2 little blue readers

www.littlebluehousebooks.com

Copyright © 2023 by Little Blue House, Mendota Heights, MN 55120. All rights reserved. No part of this book may be reproduced or utilized in any form or by any means without written permission from the publisher.

Little Blue House is distributed by North Star Editions:
sales@northstareditions.com | 888-417-0195

Produced for Little Blue House by Red Line Editorial.

Photographs ©: Shutterstock Images, cover, 4 (top), 4 (bottom), 7, 9 (top), 9 (bottom), 16, 19 (top), 19 (bottom), 21 (top), 21 (bottom), 24 (top right), 24 (bottom left), 24 (bottom right); iStockphoto, 10, 13, 15, 23, 24 (top left)

Library of Congress Control Number: 2022901945

ISBN
978-1-64619-599-2 (hardcover)
978-1-64619-626-5 (paperback)
978-1-64619-677-7 (ebook pdf)
978-1-64619-653-1 (hosted ebook)

Printed in the United States of America
Mankato, MN
082022

About the Author

Meg Gaertner enjoys reading, writing, dancing, and being outside. She lives in Minnesota.

Table of Contents

reusing

waste

Reusing

Reusing means using the same item again and again. By reusing items, people can make less waste.

People create waste by throwing items away. Some items break, but many can be fixed. Other items get old, but many can be reused.

broken

For example, people can use old clothes as cleaning rags.

They can use packing boxes to store things.

cleaning rag

packing box

donating
DONATIONS
DONATIONS

Donating

Donating items is another way to reuse items. Donating means giving to someone in need.

Sometimes items are still good, but people don’t want them.

For example, a boy does not like a shirt anymore, but the shirt is clean.

Instead of throwing the shirt away, the boy could donate it.

That way, someone else could use the shirt.

Reusable Items

Some items are made to be reused instead of thrown away. People can buy these reusable items.

A boy fills his reusable water bottle again and again.

A girl brings a reusable lunch box to school again and again.

reusable water bottle

lunch box

A family brings a reusable bag to the store.
They use that bag instead of a plastic bag.
They can use the bag again and again.

reusable bag

plastic bag

You can make less waste by reusing. Think before you throw something away. Could you use it again? Could you donate it?

Glossary

donating

reusable water bottle

reusable bag

waste

Index